Nanotechnology Research Scientists

Michael Souza

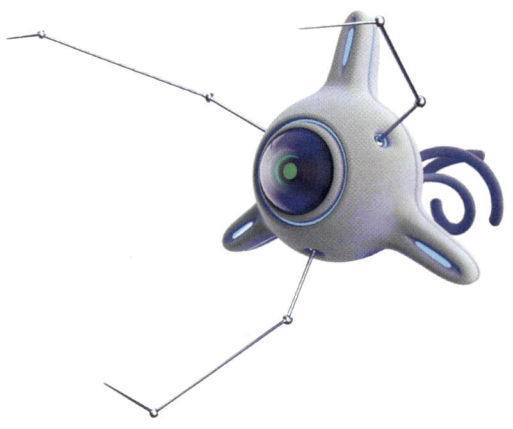

Series Editor **Casey Malarcher**

Level 4 - ❷

Nanotechnology Research Scientists

Michael Souza

Series Editor: Casey Malarcher
Acquisitions Editor: Anne Taylor
Copy Editor: Liana Robinson
Cover/Interior Design: Highline Studio

ISBN: 978-1-943980-49-9

10 9 8 7 6 5 4 3 2 1
22 21 20 19 18

Photo Credits
All photos are © Shutterstock, Inc.

Contents

What Is Nanotechnology? ·································· 4

Why Do People Study Nanotechnology? ·············· 7

Uses for Nanotechnology ························· 11

Becoming a Nanotechnologist ····················· 20

Comprehension Questions ························· 26

Glossary ································· 27

Notes ································· 31

What Is Nanotechnology?

A scientist is someone who studies or has expert knowledge of one or more of the natural or physical sciences. There are many types of scientists. Some study plants and animals. Some study space. And some scientists study ideas within mathematics and physics. Some even study things which may or may not really exist! But one of the newest and more cutting-edge types of scientists is a nanotechnology research scientist, or nanotechnologist.

Scientists at work

Nanotechnology is the moving or changing of matter at an atomic or molecular level. What does that mean? It means that a nanotechnology scientist works with things that are very, very small. Atoms and molecules are the building blocks of everything. But they are tiny. We can't see them using just our eyes. The things that nanoscientists study are even smaller than atoms. They study tiny particles called nanoparticles. The term *nano* means "one billionth" in size! That is smaller than we can see with most microscopes.

Looking at nanoparticles under a high-power microscope

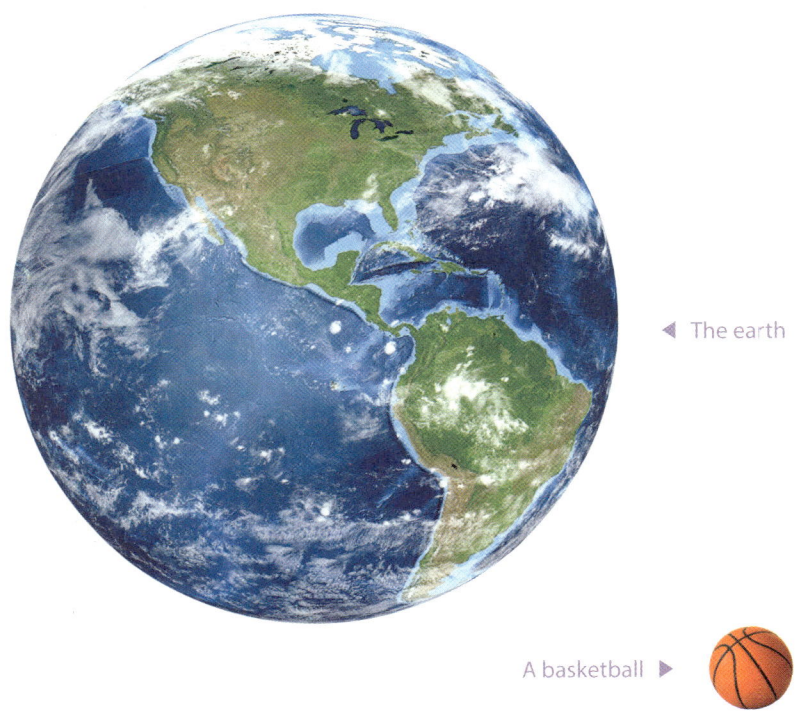

◀ The earth

A basketball ▶

In fact, nanotechnology scientists study things that are millions of times smaller than we can see with the average microscope! Imagine that you look under a microscope and see a single cell. Now imagine that cell being the size of the earth. Compared to that cell, a nanoparticle is the size of a basketball.

Why Do People Study Nanotechnology?

There are two main reasons that scientists study things this small. The first is to understand the world around us. And the second is to develop technology that we can use in our daily lives.

Researchers in a lab

Our world is controlled by certain rules, or laws, that are the same everywhere. For example, the speed of light is always 300 million meters per second. However, the laws of nature seem to be different at the nano level. For example, some objects with color have no color when they are very small. Some materials that are solid become like air when they are super-small.

An atomic model ▶
showing subatomic
particles

8

Seeking knowledge

Nanotechnology research scientists try and understand why these things happen. Why do particles behave differently based on their size? And how can we make use of these differences in our everyday lives?

It is now possible for scientists to make things using nanotechnology. Nanotechnologists can move molecules around to form certain shapes. By doing this, they can build materials with amazing uses. Did you know that the sunscreen you use was made thanks to nanotechnologists? It contains tiny nanoparticles that protect your skin from the sun.

Putting on sunscreen

Uses for Nanotechnology

Nanotechnologists work in the fields of electronics, energy production and storage, food science, and medicine, among others. There are many areas in which nanotechnology is used.

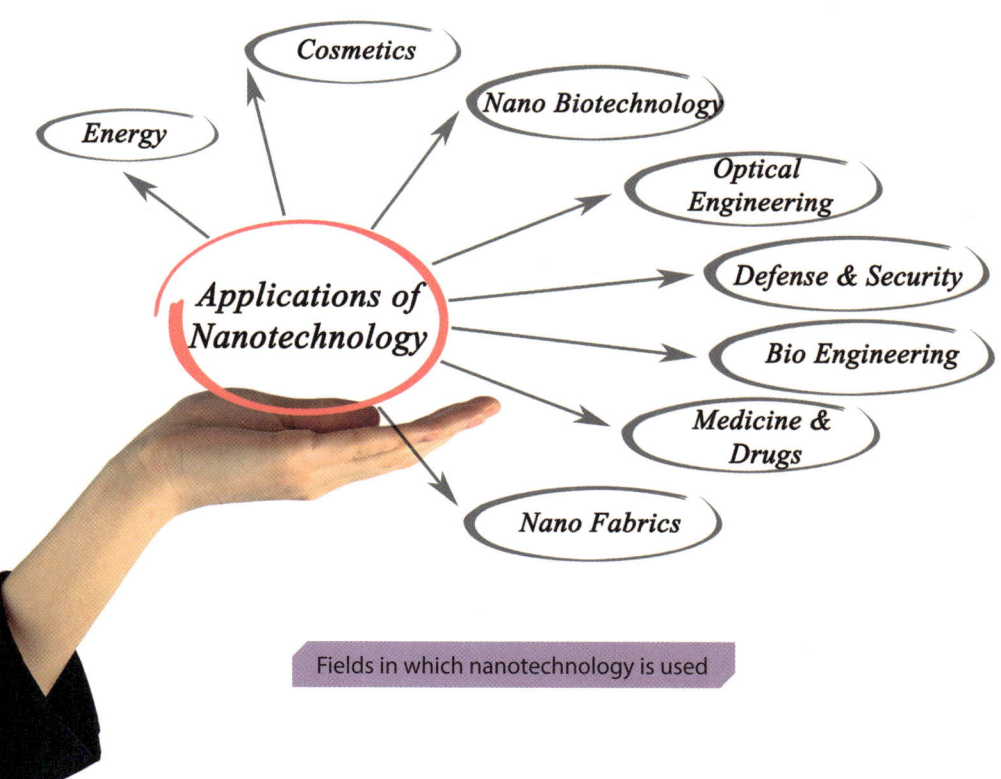

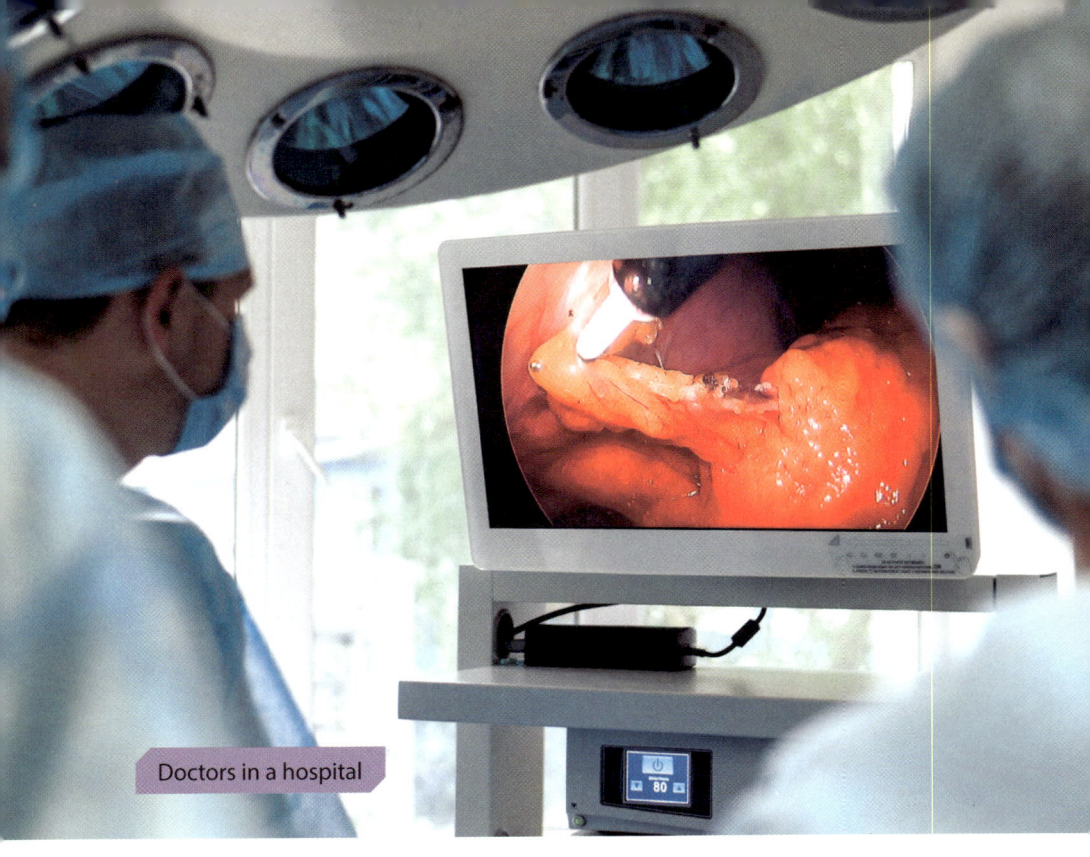

Nanotechnologists are looking at ways to use nanotechnology in the medical industry. They are already doing great things. They have found a way to attach tiny cameras to small pills that a person can swallow. These cameras can record what they see as they move through a person's body. This procedure is much less painful than similar past procedures that used larger cameras. And the new tiny cameras can go into very small spaces.

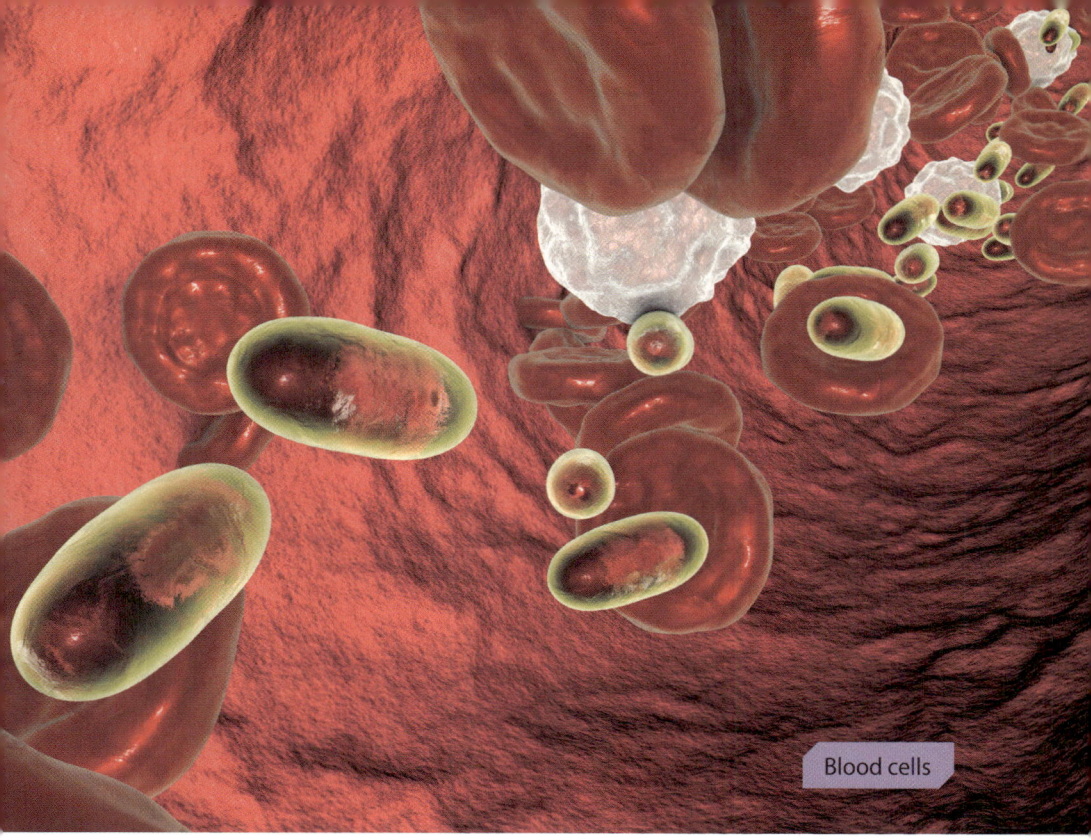

Blood cells

Nanotechnologists are also trying to make nanoparticles that stick to medicines. In the future, special nanoparticles that have been added onto medicines may be able to kill harmful cells from the inside. Scientists are also trying to use nanoparticles to grow new nerves in the human body. This will help repair damaged body parts.

In transportation, nanotechnologists are trying to develop more environmentally friendly transportation. They hope to make clean fuels for cars or special engines that don't make the air dirty.

◀ An environmentally friendly car

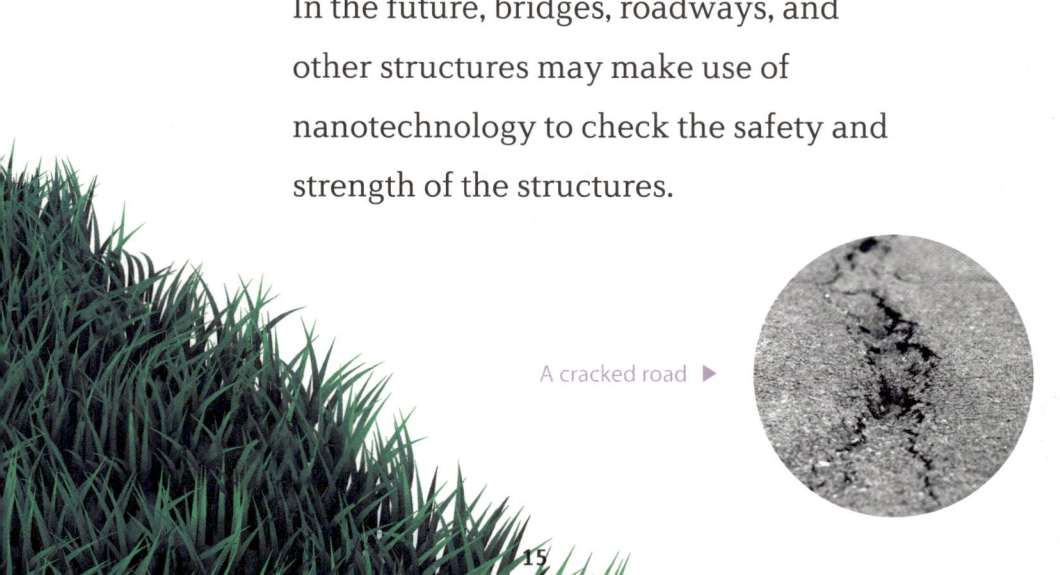

Checking a bridge over a roadway

In the future, bridges, roadways, and other structures may make use of nanotechnology to check the safety and strength of the structures.

A cracked road ▶

Nanotechnology is also being used in the computer field. Some nanotechnology scientists think that information, or data, can be stored on the nano level. It may be possible to store millions of pieces of data on a device the size of a shirt button!

A USB storage ▶
device

Saving data on a computer

Nanotechnology is important to the food industry. Nanotechnologists are working with food scientists to make safe packaging for food. Silver nanoparticles can be added to packaging to kill bacteria. Other nanoparticles stop water from entering food packaging and keep the food safe.

Shopping in a supermarket

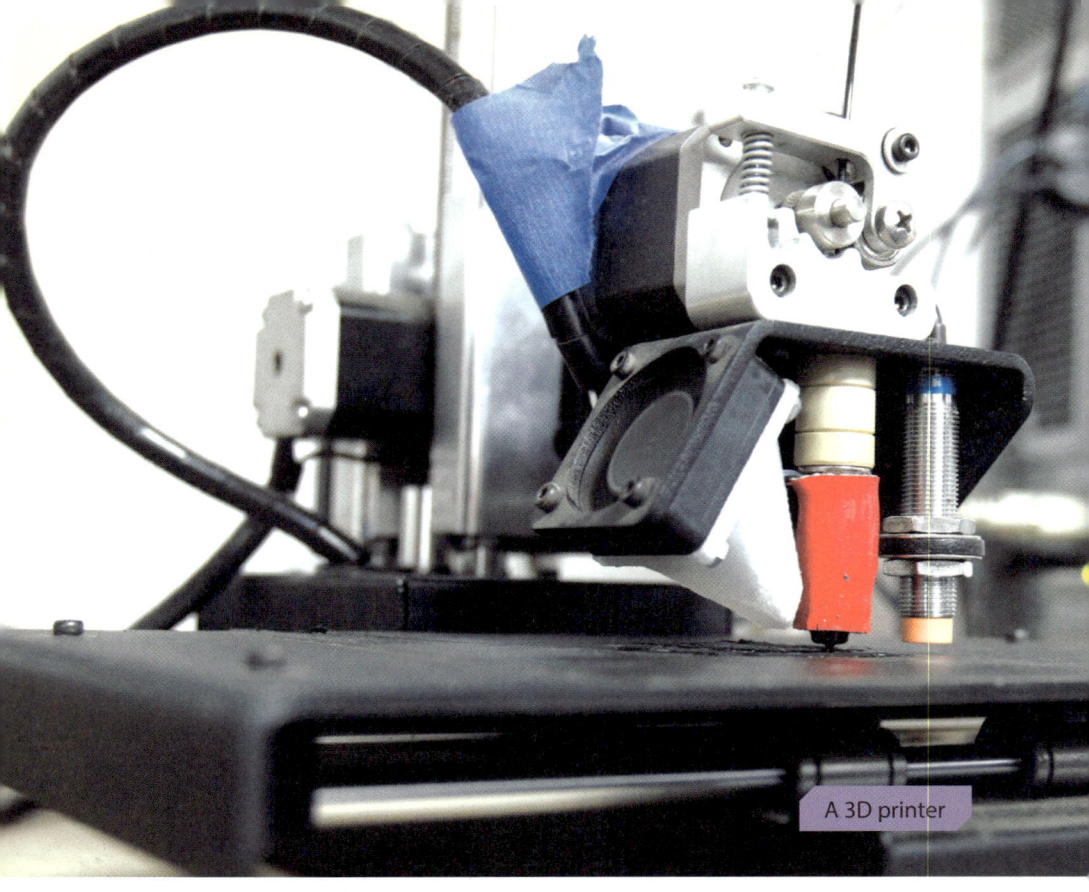

A 3D printer

Almost everyone has heard of 3D printers. They can print objects, not just images. Nanotechnologists are making "nano-printers." They can print very, very small 3D structures. For example, researchers have developed nano-printing techniques to create electronic parts with greater energy storage.

One of the newest things in nanotechnology is molecular building. This is using molecules to build things. We already use robots to build things like cars and computer parts. Scientists want to use molecules as tiny robots—nano-bots. With nano-bots, some scientists think they will be able to make everything from diamonds to food in the future.

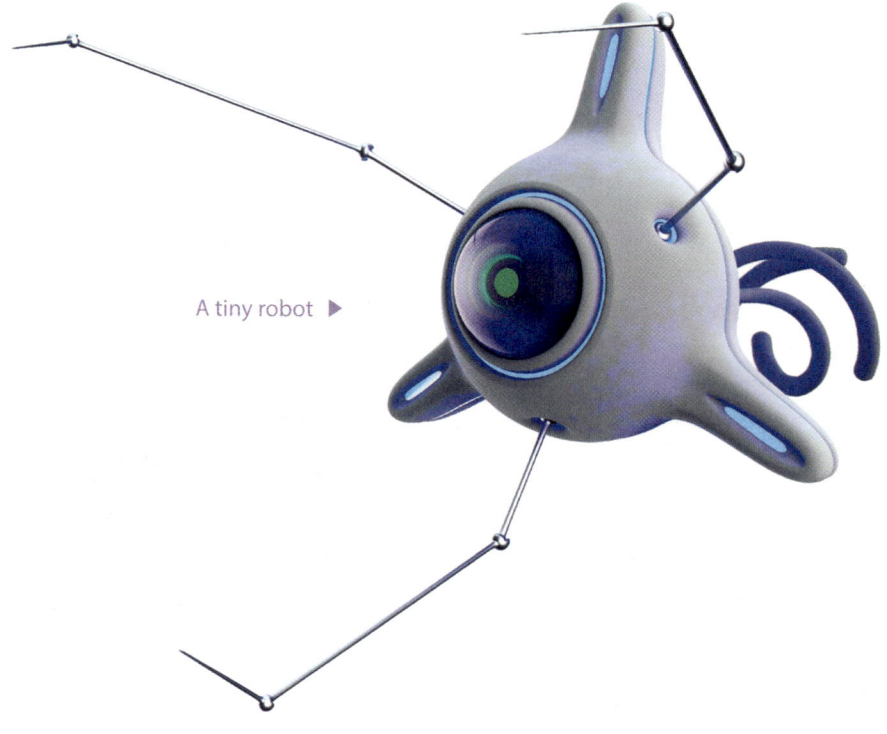

A tiny robot ▶

Becoming a Nanotechnologist

Being a nanotechnology research scientist is exciting. However, it requires a lot of education. Nearly all nanotech scientists these days hold Ph.D. degrees. They study a lot of biology, chemistry, engineering, physics, and mathematics. Many of them come from top schools. But some nanotechnologists have graduated from schools that are not famous. Being a nanotechnology research scientist is not about what school you go to. It's about how hard you study and apply yourself.

University students

Success in this field also requires that a person be good at both concrete and abstract thinking. Concrete thinking is related to thinking about facts. Abstract thinking, on the other hand, involves using one's imagination to think of ways to use those facts to solve a problem.

For example, a nanotechnology research scientist may use mathematics and chemistry (concrete thinking) to develop new ways to store computer sensors that can detect temperature changes on cloth (abstract thinking). This can make clothes that react to temperature changes to keep the wearer warmer or cooler based on the person's level of activity as well as the outside temperature.

Thinking about ▶
a problem

What is it like to be a nanotechnologist? You'll work in a laboratory, where you'll use specific scientific equipment, like high-power microscopes. You'll be expected to wear clothing like a lab coat and safety glasses.

You may work on projects with scientists in other countries, so you may need to travel for research and conferences.

Your day-to-day work may include:

- Using scientific instruments to do research
- Performing experiments to test the nanotechnology you have produced
- Checking the condition of equipment
- Using computers to understand data
- Giving speeches about your work to students and other scientists
- Planning research schedules in a laboratory
- Writing reports and articles

Listening to a speech

Students with answers

Nanotechnology research scientists can work in many places. They work in universities, medical research centers, and businesses. Research centers are often supported by private companies to study solutions to certain problems or to advance different kinds of technology.

Some nanotechnology research scientists even work from home in their home laboratories. If you love to be surrounded by science, and love solving problems, this job might be good for you.

The one thing that all nanotechnology researchers share is a curiosity about the world around them. They all want to solve problems and help people.

Nanotechnology is still fairly new, so it will continue to grow. As a nanotechnologist you could have the chance to invent something new that could change lives.

Comprehension Questions

1. What is nanotechnology?
   (a) The science of physics
   (b) The science of very small particles
   (c) The science of other countries
   (d) The science of food

2. Which is NOT mentioned as something a nanotechnologist needs?
   (a) Goggles
   (b) A lab coat
   (c) A microscope
   (d) Sunscreen

3. In what ways do some things change when they are very small?
   (a) They become cells.
   (b) They smell different.
   (c) Their color changes.
   (d) They break easily.

4. According to the text, how is nanotechnology used in the food industry?
   (a) To make safe packaging
   (b) To grow more food
   (c) To make food look good
   (d) To make bigger packaging

5. A nanotechnologist should . . .
   (a) be interested in solving problems.
   (b) have experience building robots.
   (c) make friends with many other scientists.
   (d) learn to speak a second language.

Glossary

- **abstract** (adj.) existing as thoughts in the mind

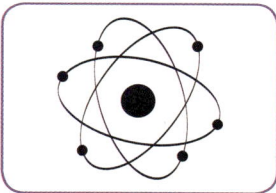

- **atom** (n.) the smallest unit of a molecule made up of protons, electrons, and neutrons

- **attach** (v.) to join one thing to another

- **billionth** (n.) one of a billion equal parts; 1/1,000,000,000

- **biology** (n.) the scientific study of living things

- **chemistry** (n.) the study of the structure of substances

- **concrete** (adj.) based on facts

- **environmentally** (adj.) in a way that affects nature

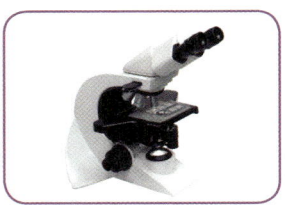

- **microscope** (n.) a piece of equipment that makes small things look bigger

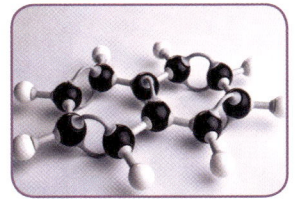

- **molecule** (n.) two or more atoms joined to each other

- **nano-** (adj.) one of a billion equal parts; involving the science of nanotechnology

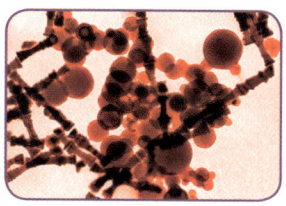

- **nanoparticle** (n.) a tiny particle whose size is measured in nanometers

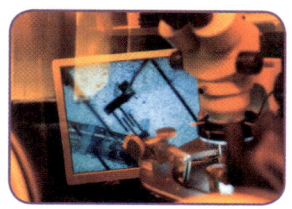

- **nanotechnology** (n.) the science of changing things at the atomic and molecular level

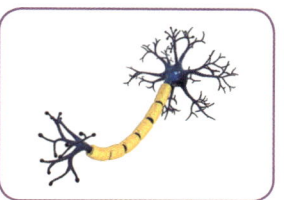

- **nerve** (n.) one of the groups of fibers in your body that carry messages to and from your brain

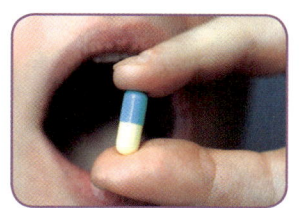

- **pill** (n.) a small, prepared amount of a medicine

- **production** (n.) the act of making something

- **physics** (n.) the science of understanding matter, energy, motion, and forces

- **sensor** (n.) a device that responds to heat, light, sound, etc. and gives off a signal

- **solid** (adj.) having a shape that does not change; not liquid or gas

- **sunscreen** (n.) a cream used to protect skin from the sun

Notes

Here are some important people related to the field of nanotechnology. Readers may enjoy researching these people to learn more about the field.

James Heath: Dr. Heath has a Ph.D. in Chemistry and teaches at Rice University. He developed a "molecular switch" that acts like the light switch in a house. Using this molecular switch, functions of molecules can be turned off and then back on again as needed!

Naomi Halas: Dr. Halas has a Ph.D. in Physics. She also teaches at Rice University and has developed over 200 different products and technological applications. Dr. Halas is helping to develop "nanoshells" that can find cancer cells, cover them up, and kill them.

Tianlong Li: Dr. Li teaches at the Harbin Institute of Technology in China. He and his colleagues built a nano-robot that can swim through a person's blood system and kill bacteria.

Morinobu Endo: Dr. Endo is a Japanese chemist and physicist who has been working with nanotechnology since the 1970s. His main area of interest focuses on the properties of carbon nanotubes.

List of Books

LEVEL 1

1. Robotics Engineers
2. Cyber Security Experts
3. Medical Scientists
4. Social Media Managers
5. Asset Managers

LEVEL 2

1. Drone Pilots
2. App Developers
3. Wearable Technology Creators
4. Computer Intelligence Engineers
5. Digital Modelers

LEVEL 3

1. IoT Marketing Specialists
2. Space Pilots
3. Water Harvesters
4. Genetic Counselors
5. Data Miners

LEVEL 4

1. Database Administrators
2. Nanotechnology Research Scientists
3. Quantum Computer Scientists
4. Agricultural Engineers
5. Intellectual Property Lawyers

"The future of the economy is in STEM. That's where the jobs of tomorrow will be."

James Brown (Executive Director of the STEM Education Coalition in Washington, D.C.)

Data from the US Bureau of Labor Statistics (BLS) support that assertion. Employment in occupations related to STEM—science, technology, engineering, and mathematics—is projected to grow to more than 9 million by 2022 [in the US alone] . . . Overall, STEM occupations are projected to grow faster than the average for all occupations.

from *STEM 101: Intro to Tomorrow's Jobs* **US Bureau of Labor Statistics**